060934

688.6 Gould, Marilyn
G

Skateboards,
scooterboards &
seatboards you can
make

DATE			
APR			
MAR 2 6			
DEC 7			
FEB 8			
FEB 26			
NOV			
APR 22			
SEP 27			
JAN 22			
OCT 1			
OCT 8			

© THE BAKER & TAYLOR CO.

P 82

Skateboards, Scooterboards & Seatboards

You Can Make

Marilyn Gould & George Gould

Skateboards, Scooterboards & Seatboards
You Can Make

drawings by Loring Eutemey
photographs by Lou Jacobs, Jr.

Lothrop, Lee & Shepard Company • New York
A Division of William Morrow & Company, Inc.

Thanks to the kids of all ages
who have helped us with so much enthusiasm

3   4   5   6   7   8   9   10

Library of Congress Cataloging in Publication Data

Gould, Marilyn.
    Skateboards, scooterboards, and seatboards you can make.

    SUMMARY: Directions for building a skateboard, scooterboard, and seatboard, with information on how to ride them.
    1. Skateboards—Design and construction—Juvenile literature. 2. Scooters—Design and construction—Juvenile literature. [1. Skateboards—Design and construction. 2. Scooters—Design and construction] I. Gould, George, joint author. II. Eutemey, Loring. III. Jacobs, Lou. IV. Title.
TT174.5.S35G68   688.6   76-52919
ISBN 0-688-41794-9   ISBN 0-688-51794-3 lib. bdg.

To Dad and Grampa
who started our wheels turning
and our dear Edie
who was always so proud

Contents

Skateboards, Scooterboards & Seatboards
You Can Make

Put Yourself on Wheels!

How would you like to put yourself on wheels? Imagine flat tracking on a custom skateboard, making slalom turns on a scooterboard, or coasting along on a seatboard, all specially made by you. This book will show you how to make and ride a skateboard, scooterboard, and seatboard. They're easy to make and lots of fun to ride.

Start with the skateboard. It's the easiest. There are only two parts to put together, the board or blank (as skateboarders call it) and the wheels. Some people buy a skateboard already made or a blank already cut and drilled, but you can have fun and save money by making your own board and cutting out your own blank. When you make it yourself, you can decide on the exact shape, size, and wheels you want; you can finish the board so it really is your own custom job.

To make a neat scooterboard, all you have to do is make a longer skateboard, move the wheels farther apart, and add a post and handle. Even if you're not a really skilled "sidewalk surfer," you'll love this scooterboard. The handle gives you something to hold onto and makes it easier to practice shifting your weight and making turns.

13

After you've made the skateboard and the scooter-board, you'll be expert enough to make and ride the seatboard. The seatboard is just like the scooterboard except that the blank is longer, the post is shorter, and it has a raised seat. You can sit on it, kneel on it, or raise and lower yourself on the seat like a motocross rider. You turn by holding onto the post with the palms of your hands and shifting the lower part of your body. You stop yourself by putting your feet on the ground.

Before you start, read the general instructions in "Getting Started" so you'll know what you're doing. Then read all the directions for the project you're working on. Make sure you have all the tools and supplies you'll need. Then go back and follow each step, one by one. These "wheels" are for everyone, girls as well as boys, so turn the page and get *your* wheels turning.

Getting Started

A PLACE TO WORK

The first step is to find a good place to work. The garage, the basement, or the backyard may be better than your bedroom. Look for a place where:

- there is enough room to spread out;
- you can saw, hammer, paint, and make a mess;
- your things will be safe and not in other people's way.

When you have finished for the day, it's a good idea to straighten up. Put your tools together. Throw out the scraps you don't need. Be sure your work will not get stepped on. Then you will be ready to start again the next day.

TOOLS YOU NEED

Check your tools before you start to work. You may need to buy or borrow some. If you borrow, be sure to return them.

These are the tools you will need for all three projects:

- a hand saw (a small saw like a 22-inch is easier to handle);

- a hammer (one with claws helps to pull nails out if you make a mistake);

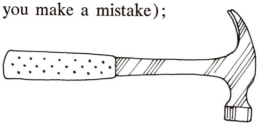

- a screwdriver;

- an eggbeater hand drill with a ⅛-inch bit and a ¼-inch bit;

- a yardstick (a metal one is best because it's more accurate);

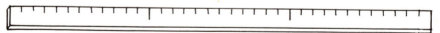

- a sharp pencil with an eraser;

- a combined rasp and file;

- an 8-inch adjustable wrench (to tighten the nuts of the machine screws);

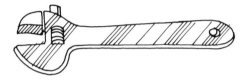

- about four sheets of sandpaper for each project (two medium for rough sanding and two fine for smooth sanding);

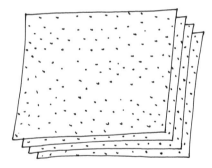

- a 4-inch C-clamp (this can be very helpful to hold things in place).

TIPS ON USING TOOLS

When you measure:
- Use a yardstick, don't guess.
- Measure, make a pencil mark, and measure again to be sure.
- Always use a yardstick to draw a straight line.

When you saw:

Saws are big and heavy. Sometimes it's hard to get them to saw. They make jagged marks, but they don't really cut. It's a good idea to practice sawing on a piece of scrap wood before you start to saw your projects. It's also a good idea to have an older person help you.

Here are some other tips that will help you saw clean straight edges:
- Use a sharp saw.
- Draw a straight dark line on the wood so you can follow it with the saw.
- Keep the piece of wood from moving. If you have a C-clamp, you can clamp the wood to a bench or saw-horse. If you have a vise, you can use it to hold the wood. You can also kneel on the wood or steady it with one hand, but be sure you saw the wood and not yourself.
- To start the cut, pull the saw back three or four times.
- Now saw back and forth with long, slow, easy strokes.
- Slow down near the end, and saw very lightly so the end of the wood doesn't break off before you have sawed it off smoothly.

Keep your thumb and fingers away from the saw's teeth.

18

When you hammer:

- Hold the hammer at the end of the handle.
- To start the nail, hold the nail wtih your fingers and give it a few light taps.
- Take your fingers away and hit the nail a little harder.
- Try to hit the nail right on the head so it will go in straight.
- If it starts to slant a little, tap it on the side to straighten it up. If it bends over, pull it out with the hammer claw and try again.
- Keep your eye on the nail and the hammer away from your fingers.

When you drill a hole:

- Just as when you saw, the wood should not move when you drill. Hold it down with a C-clamp, a vise, or your knee.
- Make a starting hole by hammering a nail partway in. Then take the nail out; the starting hole will guide your drill.
- Try to keep the drill from wobbling.

When you rasp and file:

A combined rasp and file has a rasp on one side and a file on the other. The rougher side is the rasp. It rounds off corners. The smoother side is the file. It smooths the rough edges of wood where you have sawed. These tools are easy to use, and they save sanding time.

- You can hold the rasp or file on one end with one hand or both ends with two hands.

- Use a downward motion "across the grain." The grain of wood refers to the direction of the wood fibers. You can work "with the grain" (the direction the tree grew) or "across the grain" (cutting across the fibers).

When you sand:

Sanding takes time and lots of energy, but it makes your work look terrific, and it also prevents splinters.

- You'll find it easier to sand if you wrap your sandpaper around a small block of wood.
- Use long strokes, sanding with the grain of the wood. If you sand across the grain, you will make scratchy marks.
- Start with medium sandpaper to smooth out the rough places. Then use fine sandpaper and sand until the wood is really slick.

GETTING YOUR SUPPLIES

Check to see what supplies you need. You may already have some of the things or know where you can get them without buying them. You may have to go to a hardware store, a lumberyard, or a toy store for the things you don't have. In some cities, there are surf and skateboard shops that have skateboard parts.

Write a list so you won't forget anything. To save time, telephone before you go to find out which stores have the things you need. You can also find out how much they will cost.

Lumbermen and hardware salespeople have their own code. It will help you to know that:
- " means inch or inches;
- ' means foot or feet;
- x means by;
- # means number;
- cm means centimeter or centimeters (1" = 2.5 cm).

Most tools and materials you'll be using are measured in inches and feet in the United States, not in centimeters and meters. However, you may find yourself using metric measurements sometimes. Metric measurements for working with wood are given in this book. For metric tools, refer to the chart on pages 72-73 to find the nearest equivalents.

The following facts will help you tell lumber and hardware salespeople exactly what you need for your projects.
- A two-by-four (2 x 4) is a piece of wood, usually pine, that is cut roughly 2 inches thick by 4 inches wide. But because the wood you buy is cut smoothly in a mill, a 2 x 4 is usually $1\frac{1}{2}$ inches thick and $3\frac{1}{2}$ inches wide.
- A 30" 2 x 4 is a piece of pine 30 inches long, $3\frac{1}{2}$ inches wide, and $1\frac{1}{2}$ inches thick.
- A one-by-one (1 x 1) is a piece of wood roughly cut 1 inch thick and 1 inch wide, smoothly cut $\frac{3}{4}$ inch thick and $\frac{3}{4}$ inch wide.
- A 12" 1 x 1 is a piece of pine 12 inches long, $\frac{3}{4}$ inch wide, and $\frac{3}{4}$ inch thick.
- Wood screws are usually used for wood and have a pointed end. They are numbered according to their thickness: the larger the number, the thicker the screw. Their length is measured in inches.

21

- A ¾″ #10 is a wood screw ¾ inch long; it is thicker than a ¾″ #8.
- A machine screw has a flat end and needs a nut. The length and diameter of machine screws are measured in inches. A 2½″, ³⁄₁₆″ machine screw is 2½ inches long and ³⁄₁₆ inch in diameter.
- There are many different kinds of nails, such as finishing nails for furniture and box nails for light construction. You should use common nails which are for general construction. Their size is measured in inches (a 3-inch nail) or in "pennies" (a 10-penny or 10d nail).

CHOOSING AND DECIDING

The best part of making something yourself is that you can choose the materials and size you want, and you can make it just right for you.

First, you need to decide what kind of wood you want to use for the board or blank. Pine is a soft wood. It's easier to saw and drill than a hardwood like mahogany or ash. It's also cheaper, but it won't make a very strong board and it won't be as safe because the screws can pull out more easily. If you decide to use pine, be sure it is ¾ inch thick. The grain should go the length of the board, not across the board (Figure 1).

Mahogany is the easiest hardwood to work with, and it will make a strong, safe board for all three projects. You can use ash or birch if you want a harder wood, but it will also be more difficult to saw and drill.

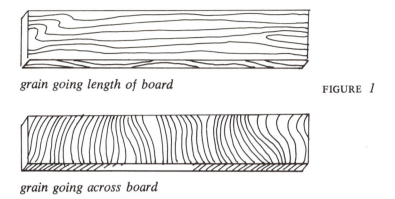

grain going length of board

FIGURE *1*

grain going across board

Many hardware stores and lumberyards only carry pine. Some carry mahogany. You may have to go to a special lumberyard to find other kinds of hardwood. They will cut the piece you choose to the exact size you need.

Next, you need to decide on the kind of wheels and trucks you want. A truck is a unit that holds the wheels. It has a baseplate that screws onto the blank, cushions to soften the ride, axles to hold the wheels, and an action bolt and nut to adjust how the board turns (see Figure 2).

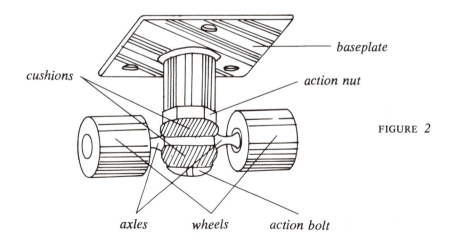

baseplate

cushions

action nut

FIGURE *2*

axles *wheels* *action bolt*

There are all kinds of trucks. Some cost much more than others. Unless you want a truck for special stunts, you don't need a very expensive one—but be sure it's safe and won't break.

There are also many different kinds of wheels. The first skateboards made had clay wheels. Clay wheels are the cheapest to use, but they're not very good and not very safe. They make noise; they hit bumps hard; and they skid and wear out easily.

Urethane wheels are easier, safer, and quieter to ride. Companies that make skateboards and skateboard wheels are constantly coming out with new and more expensive ones. They may be softer for a smooth ride, harder for a fast ride, wider for speed, or narrower for doing stunts. They may cost a lot or a little.

The average-priced urethane wheels would be a good choice for your skateboard and scooterboard. Later on, if you want to, you can replace the first wheels with more expensive ones. For your seatboard, use the least expensive urethane wheels.

You can decide on the size board you want. Skateboards are usually 24 inches long for tricks, 27 inches long for slalom, and 30 inches long for downhill. Some sidewalk surfers like their boards to be 36 inches or even longer for speed and for the feeling of riding a surfboard.

Boards can be different shapes, too. Pin tails, swallow tails, and diamond tails are a few; but most of the best boards have blunt noses and rounded tails.

The directions in this book are for a 24-inch (60-cm) skateboard, a 27-inch (67.5-cm) scooterboard, and a 30-

inch (75-cm) seatboard with rounded noses and tails. They are the easiest sizes and shapes to make and ride. If you want, you can change the design to suit yourself. Just be sure the wheels are centered on the blank and are between 2 to 3 inches (5-7.5 cm) from the ends.

Skateboards

- A piece of wood 24 inches long, 6 inches wide, and ¾ inch thick (24″ x 6″ x ¾″ or 60 cm x 15 cm x 2 cm).

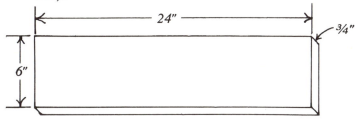

- Two pairs of skateboard wheels on trucks (this means you want wheel units, not just wheels).

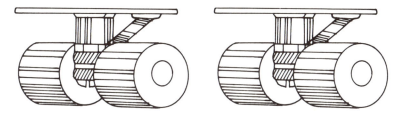

- About 10 wood screws (¾″ #10).

MAKING YOUR SKATEBOARD

1. With a yardstick, measure and draw a line down the center of the piece of wood. If your wood is exactly 6 inches (15 cm) wide, the line will be 3 inches (7.5 cm) from each side (Figure 1).

2. Now measure and mark 2 inches (5 cm) along each side from each corner (Figure 2).

3. Draw a line with your yardstick across each corner (Figure 3).

4. Saw off each corner along the lines you drew (Figure 4).

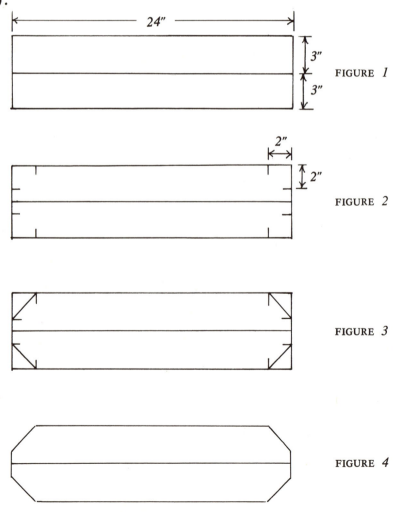

24"

3"
3"
FIGURE 1

2"
2"
FIGURE 2

FIGURE 3

FIGURE 4

5. Measure 2¼ inches (5.5 cm) from each end along the center line. Mark with lines at right angles to the center line (Figure 5).

6. Use the rasp, file, and sandpaper to round off the corners and the edges (Figure 6). (Be careful not to sand off the center line or the two marks on the center line.)

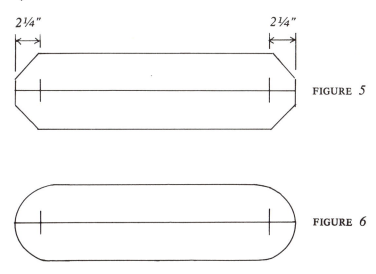

FIGURE 5

FIGURE 6

7. Work with one pair of wheels at a time. Put one pair of wheels inside one of the 2¼-inch (5.5-cm) marks. Center the baseplate of the truck on the center line with the action bolt facing in (Figure 7). Be very sure that the center of the truck is exactly on the center line.

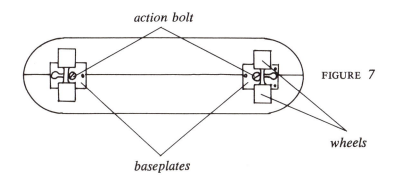

action bolt

FIGURE 7

wheels

baseplates

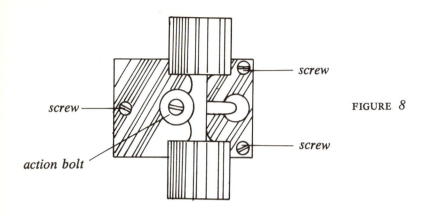

screw

screw

FIGURE *8*

action bolt

screw

8. Hold the truck firmly in position. With a pencil, mark where the screws go (Figure 8).

9. Take the trucks off and drill a hole at each mark, using a starting hole and a ⅛-inch bit (see page 19).

10. Fit the baseplate of the truck back on the holes you drilled and screw it on.

11. Do the same thing with the other pair of wheels at the other 2¼-inch (5.5-cm) mark.

FINISHING YOUR SKATEBOARD AND KEEPING IT WORKING

There are many ways to finish your board. Lacquer, varnish, or paint will protect the board and keep it clean.

To make your board less slippery, you can use grip or non-skid tape that is sold at hardware, boat, or surf and skateboard stores. It looks good to write your name or initials; or you can make your own personal design (called a logo) with non-skid tape.

Another way to put "grip" on your board is to sprinkle a little sand on the top when the lacquer, varnish, or paint is wet.

You can decide on the finish that is best for you. Maybe you want to leave the board smooth and natural.

It's important to keep your board in good working condition. The surface of the blank, the trucks, and the wheels should be kept clean. If anything gets stuck in the wheels, they can jam and cause you to fall.

If you use tape on the blank, make sure it is firmly attached so your foot won't get stuck when you want to get off the board.

All screws should be tight; check them now and then.

Your wheels will last a long time, but if they get worn out, you can replace them without buying new trucks.

RIDING YOUR SKATEBOARD

Getting the feel of your board

Learning to ride a skateboard takes time and practice. It's very much like learning to ride a bicycle. You need balance and nerve while you're in motion. It's very hard to stand on a board when it isn't moving. If you want to see how your board feels before you learn to ride it, put it on a flat surface and hold onto someone or something. With support you can see how your feet fit on it, how it wobbles from side to side, and how it tips in front and back. When you are first learning, you may not want your board too wobbly. To make it wobble less, loosen

the action nut on the truck (see page 23), screw the
bolt farther in, and retighten the nut.

Learning to fall

Before trying to ride your board, it's wise to learn how
to fall. In almost all sports you need to know how to
fall without hurting yourself. Hot riders spend a lot of
time learning to fall properly. When you practice, try to:
- lower your body by bending your knees;
- make yourself into a ball (protecting your head,
  arms, elbows, and hands);
- roll several times after the fall;
- relax and be flexible;
- keep from reaching out and stopping your fall with
  your hands;
- think about falling properly so you don't panic.

Flat tracking

Now that you have made your board, stepped on it while
holding onto someone, and practiced falling, it's time to
learn how to ride it.

Take your board to a flat, smooth, safe place like a
park, playground, or backyard. Put one foot straight on
the board behind the front wheels. Leave room for your
back foot to go behind your front foot, but don't stand
so far forward that the board tips.

Give a little push with your other foot. Don't try to
put both feet on yet. Just pedal around to get a feel of
your board in motion. This is called "flat tracking."

See which foot feels more comfortable on the front of

the board. If it's your left foot, you're "standard" or "regular footed." If it's your right foot, you're a "goofy foot." It doesn't matter which way you ride. Skateboarders do what is comfortable for them, and either way is right.

Riding down a slight incline

When you feel ready, take your board to a place that has a very slight incline and is safe from cars or pedestrians. Perhaps there is a little slope in the same playground, park, or backyard where you have been practicing your flat tracking. Driveways aren't very safe because you can easily roll into the street.

Your board should be facing down the incline. Put your first foot straight on the board behind the front wheels and give a little push with your other foot just as you have already practiced. Now put your pushing foot behind your front foot across, or perpendicular to, the board. At the same time, turn your front foot so it's across the board too.

Keep your knees slightly bent and lean a little toward the front or nose of the board. When you do this you're "going for it." It's natural to want to lean back, away from the hill, but just as in skiing and surfing it's wrong and will make you lose your balance.

Keep your arms loose and away from your body.

If you feel you are going to fall or if you want to get off, step off the front or side of the board. The board may scoot behind you, but you will be firmly on the ground.

Keep trying this until you feel relaxed and comfortable.

Turning

When you feel you have control, you can try a turn. To learn to make turns, you must think of your body as two halves: the top half above your waist and the bottom half below your waist.

Your top half always faces down the incline or the "fall line" of the hill. Your shoulders stay level, your arms are loose, and your head and eyes look straight ahead. Your bottom half is the part that maneuvers and turns the board.

To turn in the direction your toes are pointing (a front side turn), put your weight on the side of the board your toes are on. Remember, your feet are across the board. The board will tilt down a little under your toes and turn in that direction. More weight makes a sharper turn. Less weight makes a more gradual turn. Try it over and over until you've got the hang of it.

Now try a turn in the direction of your heels (a backside turn). Put weight on the side of the board your heels are on. The board will tilt slightly under your heels and turn that way.

Always keep your top half facing forward, your knees slightly bent, and your arms loose. In time you won't have to think so much about what you're doing. You'll quite naturally shift your weight and lean in the direction you want to turn.

Try riding your board slowly on a straight line; then try a curved line. Practice until you have it just right. You'll be a real "wonder rider" when you can "weight and unweight" easily.

When you've mastered that, you can try a slow slalom

course. Set up objects or markers in a line with spaces in between. Try weaving or "wedeling" (pronounced vaydelling) around them.

Take things one step at a time. You will be less likely to fall or hurt yourself if you don't try anything new until you really feel ready.

TAKING CARE OF YOURSELF

There are many people who don't like skateboards. They think they're noisy and dangerous, and they can see no fun in them.

Police departments worry about skateboarders who have no consideration for drivers or pedestrians, who dart out into the street or speed down steep hills. They worry because such riders can cause very serious accidents.

Safety departments worry, too, about "road rash riders." "Road rash" means the bruises that skateboarders get from scraping along cement. "Road rash riders" are those who ride in cement drains, empty swimming pools, and other unsafe places. It's foolish to ride where it's unsafe. There have been fatal accidents.

You can have lots of fun with your skateboard and be a real "hotdogger" without speeding down dangerous hills or being a daredevil. Most skateboarders think you have the "smarts" if you follow safety precautions like these:

- Be sure your board is in perfect working condition with the screws and nuts tight and the wheels clean.
- Check the surface you're riding on to see how even it is, and whether it has holes, nicks, or pebbles that

might cause your board to "slide out" from under you.

- Never ride on wet or even damp pavement. Your wheels will slip.
- Wear shoes (sneakers grip best) and protective clothing at all times. Even though many skateboarders are pictured barefooted and not wearing much clothing, the champions wear shoes, long sleeves, gloves, elbow pads, and helmets. At skateboard parks, elbow pads, knee pads, and helmets are required. Designers are trying to make a really good skateboard suit that will be cool but properly padded to cut down on bruises and road rashes.
- Do road rash riding only at skateboard parks that have been safely set up for riding cement walls.
- If you're someplace that's dangerous to you or a pedestrian, or if skateboards are not allowed, tuck the board under your arm and walk.

When you ride your board properly, take extra care of how, when, and where you ride, and use every safety precaution, you will be not only a "sidewalk surfer" but an artist on wheels.

TIPS FOR SUPER SKATEBOARDERS

Skateboard tricks

All sorts of tricks can be done on skateboards, but they should be tried only by skillful riders who have perfect control. Here are some of them:

- wheelies—riding on the back wheels;

- nose wheelies—riding on the front wheels;
- 180's or endovers—spinning a half turn;
- 360's—spinning a complete turn;
- shoefly christies—sitting on one leg and sticking the other out straight;
- down curbies—going down curbs or bumps;
- up curbies—going up curbs or bumps;
- toe tapping—moving the front of the board from side to side;
- walking the board—walking up and down the board;
- hang 5—hanging the toes of one foot over the nose of the board;
- hang 10—hanging the toes of both feet over the nose of the board;
- hang heels—hanging the heels over the tail end of the board;
- two board tricks—one person does tricks on two boards;
- tandem tricks—two people do tricks on one board.

Then there are jumps, handstands, freestyle, and hot-dogging (any stunt that hasn't yet been tried!).

Skateboard clubs

There are two big skateboard organizations, Pro/Am Skateboard Racers Association and United States Skateboard Association. They put on national championships and exhibitions. Most of the big competitions have been held in New York and California, but some city parks and recreation departments in other areas are starting to have local events.

Skateboard organizations are very interested in building skateboard parks, setting aside safe places to ride skateboards, and encouraging safe riding and the use of safety equipment.

Skateboard language

Skateboarding has many of its own words and expressions that you may not find in a dictionary. Most of them come from surfing and skiing. Here are some common terms you may hear:

- blank—the board part of the skateboard;
- downhill—a speed race;
- flat tracking or pedaling—pushing along and gliding on a flat surface;
- flex—the bending or "give" of the board;
- freestylers—skateboarders who do tricks;
- goofy foot—riding with the right foot forward;
- hotdoggers—same as freestylers;
- hot riders or wonder rollers—topnotch skateboarders;
- kicking out—getting off the board by kicking it out from under;
- land skiing—a new sport in which wheels are put on snow skis so skiing can be done on "hard hills";
- pumping—going faster and slower by pushing up and down on the board, or pumping your arms by moving them up and down;
- road rashes, raspberries, and burgers—bruises;
- road rash riders—skateboarders who skate in cement bowls or empty swimming pools;
- sidewalk or road surfers—anyone who skateboards;
- skate quiver—one or more skateboards that someone owns;

- slalom—riding around markers or cones;
- sliding out—the wheels slipping out from under, usually causing a fall;
- snap or punch—the way the board comes back or reacts after it has flexed;
- standard or regular foot—riding with the left foot forward.

If you learn all these terms, no one will be able to understand what you're talking about—except other "wonder rollers!"

Scooterboards

WHAT YOU NEED

- A piece of wood 27 inches long, 6 inches wide, and ¾ inch thick (27" x 6" x ¾" or 67.5 cm x 15 cm x 2 cm). See page 22 for help in deciding what kind of wood to use.

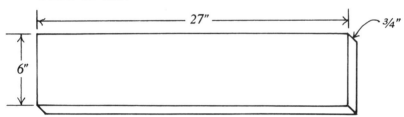

- A 2 x 4 that is 30 inches long (75 cm).

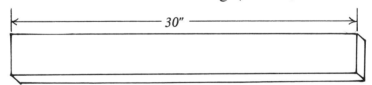

- A 1 x 1 that is 12 inches long (30 cm).

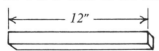

- A U-joist hanger for a 2 x 4 (you can buy this in a lumberyard or hardware store that sells lumber).

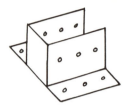

- Two bicycle hand grips.

- Two pairs of skateboard wheels on trucks (to find out more about wheels and trucks, look at page 23).

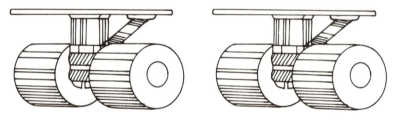

- About 10 wood screws (¾" #10).

- About 5 common nails (3" or 10d).

- Two machine screws (2½", ³⁄₁₆"), with two nuts and four washers to fit them.

MAKING YOUR SCOOTERBOARD

1. With a yardstick, measure and draw a line down the center of both the top and bottom of the wood. Write *top* on one side and *bottom* on the other (Figure 1).

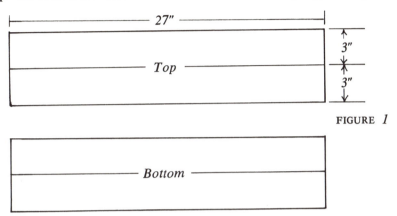

FIGURE *1*

2. On one side, measure and mark 2 inches (5 cm) along each side from each corner (Figure 2).
3. Draw a line with your yardstick across each corner (Figure 3).
4. Saw off each corner along the lines you drew (Figure 4).

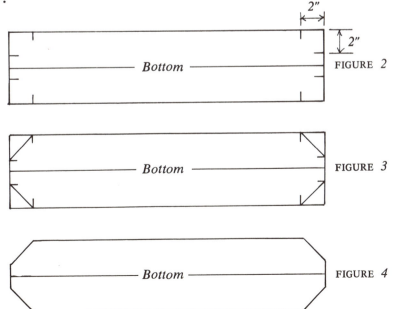

5. On the *bottom,* measure and mark 2¼ inches (5.5 cm) from each end along the center line (Figure 5).

6. Use the rasp, file, and sandpaper to round off the corners and the edges (Figure 6). (Be careful not to sand off the center lines or the marks you have made on the bottom center line.)

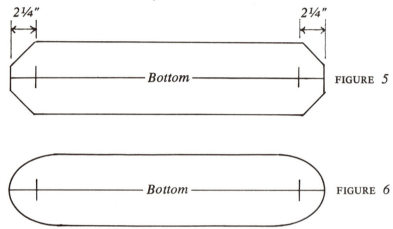

FIGURE 5

FIGURE 6

7. On the *top,* measure and mark 3 inches (7.5 cm) from one end along the center line (Figure 7).

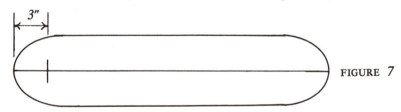

FIGURE 7

8. On the 2 x 4, measure and draw a line down the center of the length of the wood (Figure 8). (A 2 x 4 is usually 3½ inches wide, so the center will be 1¾ inches from each edge.)

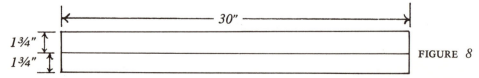

FIGURE 8

9. Set the 2 x 4 on the 3-inch (7.5-cm) mark you made on the *top* of the blank. Match up the center line of the 2 x 4 with the center line of the blank (Figure 9).

10. Turn everything over so you can nail through the *bottom* of the blank into the 2 x 4 (Figure 10). Use a vise or perhaps a friend to help you hold the two pieces of wood together and keep everything centered. (If you want the base of the handle to be attached more firmly to the blank of the scooterboard, drill holes with the ⅛-inch bit and use screws instead of nails.)

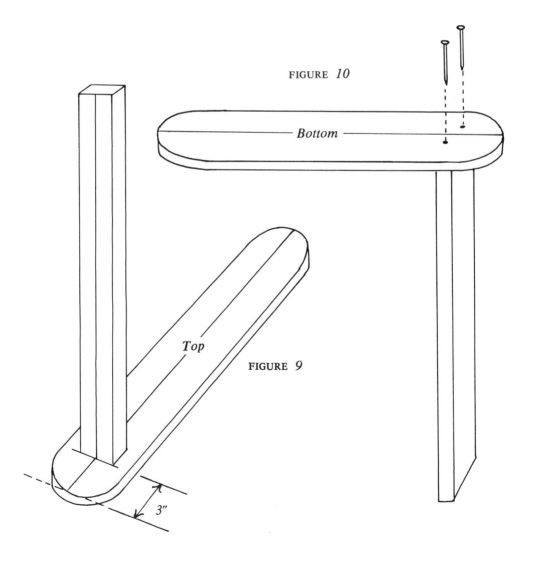

FIGURE *10*

Bottom

Top

FIGURE *9*

3"

11. Turn the scooterboard over so the *top* is facing up. Slip the U-joist hanger around the 2 x 4 so the wider plate of the U-joist hanger is on the 2 x 4 and the narrower plate is on the blank (Figure 11).

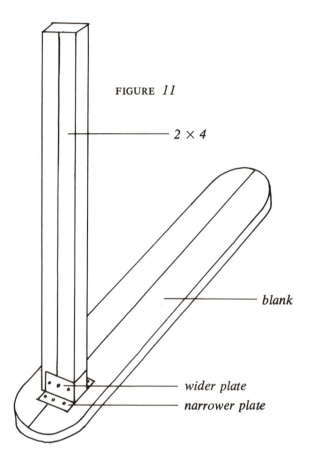

FIGURE *11*

2 × 4

blank

wider plate

narrower plate

12. Put a C-clamp around the U-joist hanger to hold it in place, but don't cover the end holes of the plates.

Drill holes through all the holes of the U-joist hanger's plates. Screw the U-joist hanger onto the 2 x 4 and the blank (Figure 12).

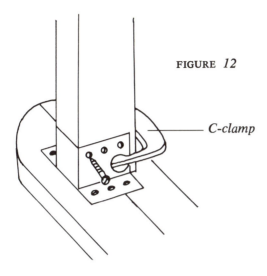

FIGURE *12*

C-clamp

13. On the 1 x 1, measure and mark the center with a pencil (Figure 13). If the 1 x 1 is exactly 12 inches (30 cm) long, the center will be 6 inches (15 cm) from each end.

14. Now measure and mark 1 inch (2.5 cm) from each side of the center for your screw holes (Figure 14).

15. This is the best time to push the bicycle hand grips all the way onto the ends of the 1 x 1 (Figure 15).

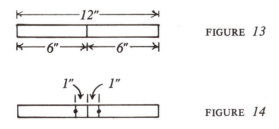

FIGURE *13*

FIGURE *14*

FIGURE *15*

16. Put the 1 x 1 on the back of the top edge of the 2 x 4, and clamp the two pieces together. Make sure the center lines are matched up.

At each of the screw hole marks, drill a hole through both the 1 x 1 and the 2 x 4, using a ¼-inch drill bit (Figure 16).

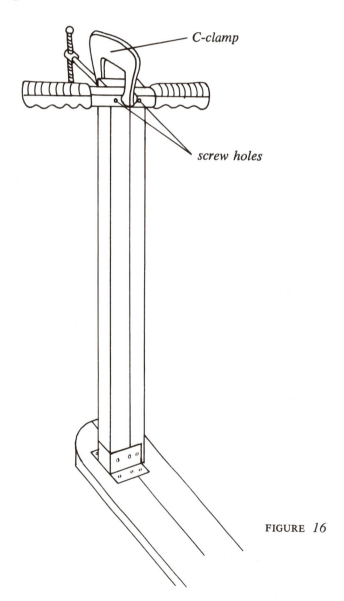

C-clamp

screw holes

FIGURE *16*

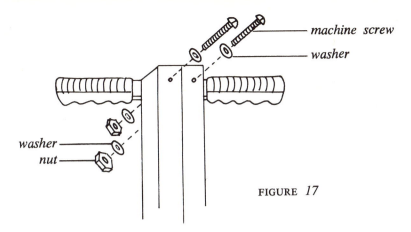

machine screw

washer

washer

nut

FIGURE *17*

17. Put a washer on each of the machine screws. Put the machine screws through the 1 x 1 and 2 x 4 (Figure 17). Then put another washer and a nut on each screw, and tighten with a wrench.

18. Turn the scooterboard over. Work with one pair of wheels at a time. Put one pair of wheels inside one of the 2¼-inch (5.5-cm) marks you made on the bottom. Center the baseplate of the truck on the center line with the action bolt facing in (Figure 18).

19. Mark where the screws go. Then take the truck off and drill a hole at each mark, using a starting hole and a ⅛-inch bit. Fit the truck back on the holes you drilled, and screw it on (Figure 19).

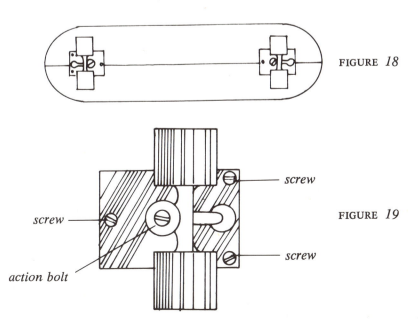

FIGURE *18*

screw

screw

FIGURE *19*

screw

action bolt

Do the same thing with the other pair of wheels at the other 2¼-inch (5.5-cm) mark.

20. Use the rasp, file, and sandpaper to round off the edges of the post and handle. Smooth all the rough spots.

FINISHING YOUR SCOOTERBOARD

You can finish your scooterboard just as you did your skateboard. Lacquer, varnish, or paint will give it a clean look and also protect it.

You can make designs on the blank of the scooterboard with non-skid tape or bathtub stickers to make it less slippery. Decals or stickers in the shape of footprints are fun to put on your scooterboard in the position that you like to ride (Figure 20).

Another nice finishing touch is to get a rubber floor mat, cut and fit it to the blank, and glue it on with white glue. They are sold at hardware and variety stores (Figure 21).

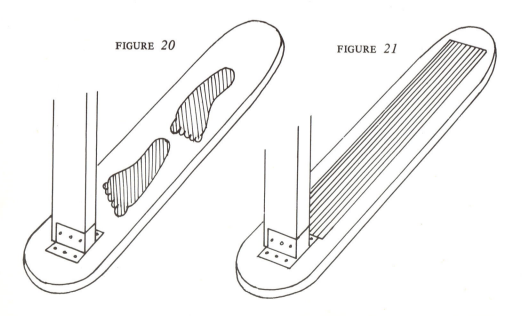

FIGURE *20*   FIGURE *21*

RIDING YOUR SCOOTERBOARD

This scooterboard is really a skateboard with a handle. The handle gives you something to hold onto and makes it easier to balance, turn, and shift your weight.

There are three ways to ride the scooterboard. You can use the flat track position, the surfer position, or the skiing position. The flat track position is good for beginners and for getting places. The surfer position is the way most skateboarders and surfers ride. The handle will help you practice turns in the surfer position with your feet across the board. The skiing position will give you the feeling of snow skiing. You can practice parallel turns.

Flat track position

Take your scooterboard to a flat safe place. Put your front foot on the board (left or right, whichever is comfortable) right behind the base of the handle. Push the scooterboard with your other foot.

Now put your back foot on the back of the scooterboard perpendicular to the board. (Your front foot faces forward.) Keep both hands on the handle, but your weight should be on your feet.

To turn, lean a little in the direction you want to turn. The handle will help you so you don't have to lean too much.

After you've got the feel of the board, try to turn using only your body weight on your feet. Keep holding onto the handle, but put as little weight on it as possible. Be careful not to overdo your turn. With the handle, your board will turn quickly and could slide out from under you.

51

Surfer position

When you feel you know how to flat track, you are ready to try the position that most skateboarders and surfers use.

Start by flat tracking. Then turn your front foot so it is also perpendicular to the board.

Hold the handle with one hand at the very top of the 2 x 4; don't hold the cross piece. If your front foot is your left foot, hold onto the handle with your left hand. If your front foot is your right foot, hold onto the handle with your right hand.

Stretch your other hand out behind you. This is called "Quasimodo."

In time you will find you can really lean, bend your knees, and weight and unweight.

Skiing position

To practice skiing on a hard surface, put your feet parallel and together, facing forward, on the center of the board.

The two ends of the handle will act as your ski poles.

You can practice bending your knees, making turns, and pushing your weight down and up on the board. (In snow skiing this is often called the down-up-down motion.) It's used in making parallel turns.

TAKING CARE OF YOURSELF

Anything that is dangerous on a skateboard is just as dangerous on a scooterboard. Some important things to remember are:

- Choose a safe place to ride.
- Beware of traffic and steep hills.
- Keep your scooterboard in good working condition. Make sure it is rust-free with clean wheels and tight screws and nuts.
- Remember that pedestrians have the right of way and don't enjoy having scooterboard riders wedeling around them.

Seatboards

WHAT YOU NEED

- A piece of mahogany 4½ feet long, 6 inches wide, and ¾ inch thick (4½′ x 6″ x ¾″ or 135 cm x 15 cm x 2 cm).

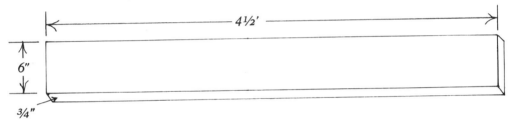

4½′

6″

¾″

- A 2 x 4, 4 feet long (120 cm).

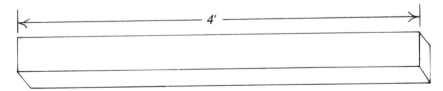

4′

- A 4 x 4 U-joist hanger.

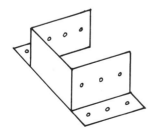

- A 2 x 4 U-joist hanger (you can buy U-joist hangers in a lumberyard or a hardware store that sells lumber.)

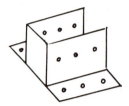

- Two pairs of skateboard wheels on trucks.

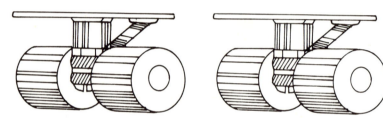

- About 30 wood screws (¾" #10).

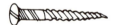

- Two wood screws (2½" #12).

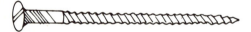

- Two machine screws (3½", ³⁄₁₆"), with two nuts and four washers to fit them.

MAKING YOUR SEATBOARD

1. Measure and saw the piece of mahogany into two pieces; make one piece 24 inches (60 cm) long and the other 30 inches (75 cm) long (Figure 1).

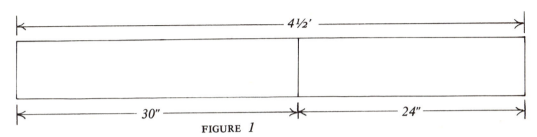

FIGURE *1*

2. Measure and draw a center line on the top and bottom of each piece (Figure 2). Write *top* and *bottom* on each piece.

FIGURE *2*

3. On the *bottom* of the 30-inch (75-cm) piece, measure and mark 2 inches (5 cm) along each side from each corner. Draw a line with your yardstick across each corner (Figure 3).

4. Saw off each corner along the lines you drew (Figure 4).

5. On the *bottom,* measure and mark 2¼ inches (5.5 cm) from each end along the center line (Figure 5).

6. Use the rasp, file, and sandpaper to round off the corners and the edges (Figure 6). (Be careful not to sand off the center line or the marks you have made on the center line.)

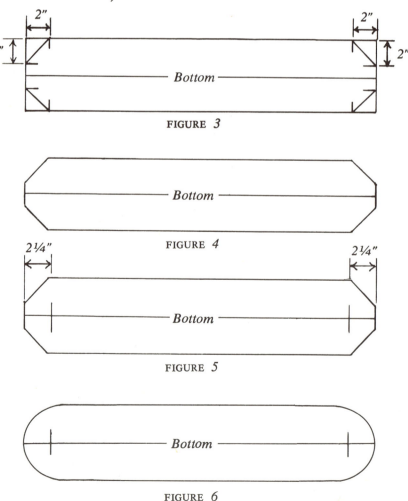

FIGURE 3

FIGURE 4

FIGURE 5

FIGURE 6

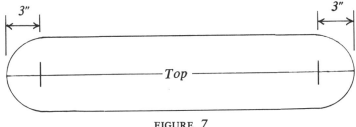

3"   3"

Top

FIGURE 7

7. On the *top*, measure and mark 3 inches (7.5 cm) from each end along the center line (Figure 7).

8. Measure and mark 1⅛ inches (2.8 cm) from the open edge of each U-joist hanger.

9. Place the 4 x 4 U-joist hanger on one 3-inch (7.5-cm) mark on the top of the blank. Line up the 1⅛-inch (2.8-cm) marks with the center line of the blank (Figure 8).

Mark on the blank where the screws go. Then take the U-joist hanger off and drill a hole at each mark, using starting holes and a ⅛-inch bit.

Fit the U-joist hanger back on the holes and screw it on with ¾" #10 wood screws.

Then do the same thing with the 2 x 4 U-joist hanger on the other 3-inch (7.5-cm) mark. Figure 8 shows how the blank should look.

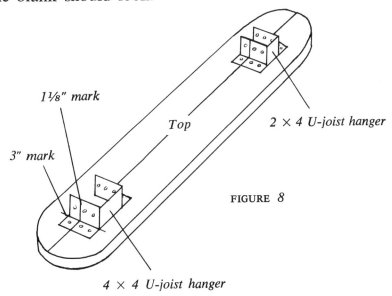

1⅛" mark

Top

2 × 4 U-joist hanger

3" mark

FIGURE 8

4 × 4 U-joist hanger

10. On the 2 x 4, measure, mark, and saw off two 12-inch (30-cm) pieces. Be very exact. You should have three pieces, two 12 inches (30 cm) long and one 24 inches (60 cm) long. See Figure 9.

11. Clamp the 24-inch (60-cm) piece and one 12-inch (30-cm) piece together with a C-clamp so the bottoms are even (Figure 10).

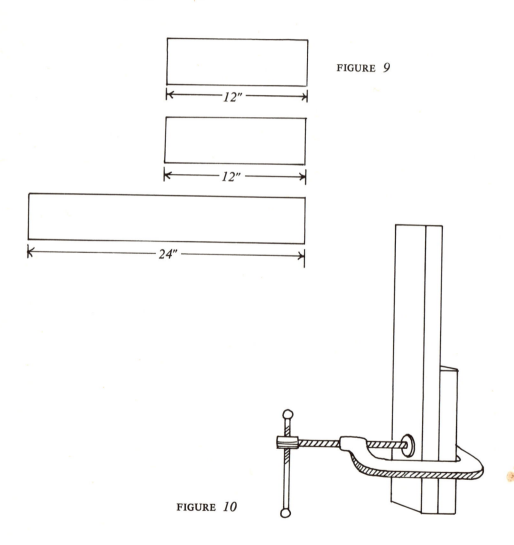

FIGURE 9

FIGURE 10

12. Measure and draw a center line on the 12-inch (30-cm) piece.

Then measure and mark 3 inches (7.5 cm) from each end of the 12-inch (30-cm) piece along the center line. (Figure 11).

13. Drill all the way through both 2 x 4's at the 3-inch (7.5-cm) marks, using a ¼-inch bit.

Put a washer on each machine screw. Put the machine screws through the holes of the 12-inch (30-cm) and 24-inch (60-cm) 2 x 4's (Figure 12). Put another washer and a nut on each screw, and tighten with a wrench.

FIGURE *11*

FIGURE *12*

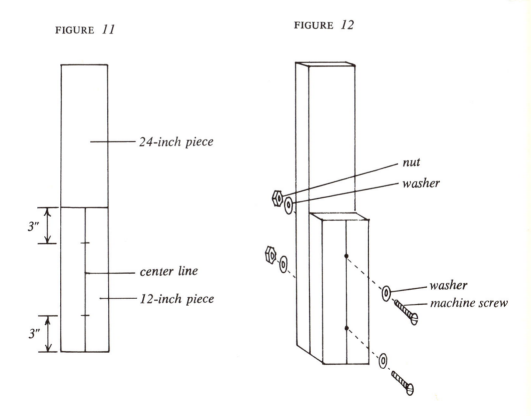

14. Put the two 2 x 4's in the 4 x 4 U-joist hanger to help hold the wood in place, but don't cover the end holes of the plates.

Drill holes through all the holes of the plates of the U-joist hanger, using a ⅛-inch bit. Then screw the U-joist hanger to the 2 x 4's with ¾″ #10 wood screws (Figure 13).

15. Put the other 12-inch (30-cm) 2 x 4 in the 2 x 4 U-joist hanger at the back of the blank. Drill holes through the holes of the plates of the U-joist hanger and screw it to the 2 x 4. Figure 14 shows the blank with both 2 x 4's attached.

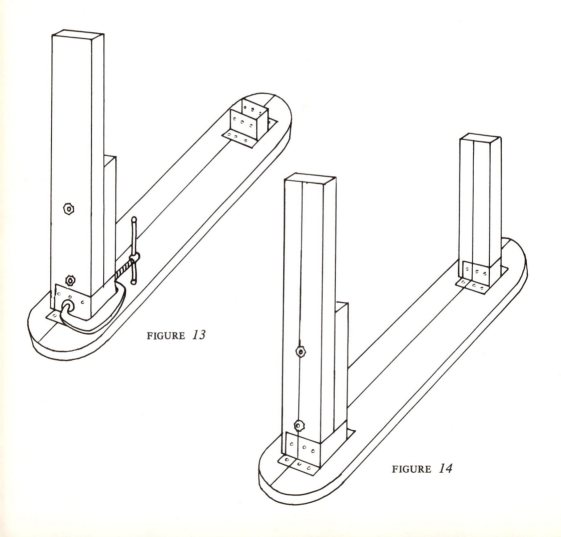

FIGURE *13*

FIGURE *14*

16. Lay the 24-inch (60-cm) piece of mahogany on the two 12-inch (30-cm) 2 x 4's. Make sure you line up all the center lines.

In the front, measure and mark 1 inch (2.5 cm) from the end along the center line. In the back, measure and mark 2½ inches (6.3 cm) from the end along the center line (Figure 15).

Drill a hole at each mark, using the ⅛-inch bit. Screw the seat to the 2 x 4's with 2½" #12 wood screws.

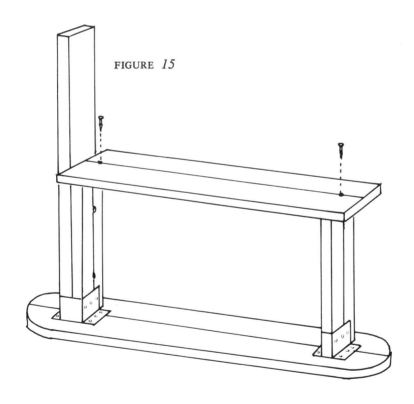

FIGURE *15*

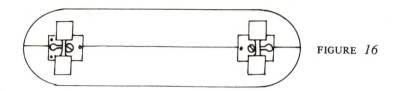

FIGURE *16*

17. Turn the seatboard over. Work with one pair of wheels at a time. Put one pair of wheels inside one of the 2¼-inch (5.5-cm) marks you made on the bottom. Center the baseplate of the truck on the center line with the action bolt facing in (Figure 16).

Mark where the screws go. Take the truck off and drill a hole at each mark, using a starting hole and a ⅛-inch bit. Fit the truck back on the holes you drilled and screw it on.

Do the same thing with the other pair of wheels at the other 2¼-inch (5.5-cm) mark.

18. Use the rasp, file, and sandpaper to round off all the edges and make the seatboard smooth.

If you think the seatboard looks unfinished, you can put on a very short handle like the one on the scooterboard, but it rides better without a handle.

FINISHING YOUR SEATBOARD

If you want your seatboard to have a more streamlined look, you can narrow down the seat so that it has more of a saddle shape. Draw a curved line from each side of the post to the side edge of the seat. Be sure both sides are the same (Figure 17). Saw off the extra wood and sand the rough edges.

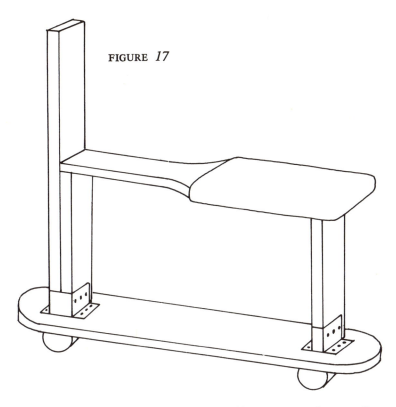

FIGURE *17*

Lacquer, varnish, or paint will protect the surface of your seatboard. Clear lacquer or varnish will make the natural grain of the wood show up better.

If you want a very hard finish, use more than one coat. Paint on the first coat. Let it dry for two or three days (in damp weather it may take even longer). When it is completely dry and not the least bit sticky, sand the surface with fine sandpaper until it is very smooth. Then use steel wool to make it even smoother. The more you rub, the smoother your surface will be. Brush off all the sandpaper and steel wool bits before you put on the second coat.

When the second coat is dry (another three days),

65

sand and steel wool the surface again. A good finisher puts on many coats, rubbing the surface smoothly between each coat. It takes a lot of time and patience, but your seatboard's surface will be beautifully slick.

An easier way to finish your wood is to wipe on an oil wood sealer. You won't get a hard or shiny finish, but the sealer will soak into the wood and protect it.

The back end of the seat of your seatboard can be padded and upholstered with foam rubber and vinyl or Naugahyde. Cut a piece of foam rubber the size of the part of the seat you want to cover. Cut a piece of material about four inches longer and wider than the part that is going to be covered. Put the foam rubber on top of the seat and then the material on top of the foam rubber. Use staples or brads on the underside of the seat to hold the material on. If you do the sides first and take your time to fold the ends and corners, you will get a neater job.

RIDING YOUR SEATBOARD

The seatboard is easy to ride, but it takes practice to learn how to turn it.

Sitting position

Start by sitting comfortably on the seat. Hold the sides of the post with the palms of your hands. Your thumbs should be on top of the post.

Put one foot on the bottom blank and push off two times with your other foot.

Rest that foot on the bottom blank and push two times

with the first foot. By flat tracking twice with each foot, you will develop a rhythm that will keep your seatboard going and keep you from getting tired.

When you coast, put both feet on the blank on either side of the post.

Kneeling position
Another way to ride the seatboard is to put one knee on the seat and flat track with the other foot.

Turning
To turn, think of your body as two halves, the top half above your waist and the bottom half below your waist. Your top half stays level and faces forward. Your bottom half shifts from side to side to maneuver the board.

To make a gradual right turn in the sitting position, bend your body at the waist. Keep your top half level and facing forward. Keep your knees together, and push them to the right. The post will go to the right and the board will tip down so it feels like the wheels on the left are leaving the ground. Keep your weight on the seat and both feet on the blank.

To make a gradual left turn, lean your knees and the post to the left, keeping the upper part of your body straight. The board will tip so it feels like the right wheels are leaving the ground.

Practice until you can turn without sliding out.

Some people can make a sharp turn before they learn to make a gradual turn, particularly those who have done motocross riding and know how to lift themselves off their seats when they take a hill or "whoop-de-doo."

To make a sharper right turn, hold onto the post in the same way. Put your feet a little farther back on the blank. Unseat yourself or lift yourself off the seat. Keep your knees together and push them to the right, tilting the board and the post to the right. Your weight will be on your feet. The top part of your body will stay straight and facing forward.

To make a sharper left turn, lift yourself a little off the seat, keep your knees together, and push them to the left. The board and post will tilt to the left.

It may take some time before you are able to maneuver your seatboard with ease, but remember how long it took to learn to ride a skateboard!

TAKING CARE OF YOURSELF

Taking care of yourself means playing it safe. Ride your seatboard in a safe place, away from traffic, where you can control the speed.

Your seatboard is a kind of vehicle, and you are the driver. Be cautious and ride where larger vehicles, like cars and bicycles, won't have to dodge you. Be a considerate driver and steer clear of smaller vehicles and those moving more slowly.

You've spent a lot of time making your seatboard and learning to ride it, so take pride in it and bring it indoors when you have finished riding it. That way it will stay clean, free of rust, and ready for another day of fun.

Make Your Own Wheels

When you make your own skateboard, scooterboard, and seatboard, you'll know how to put them together and take them apart. You'll be able to repair them and replace parts. You might even invent something new.

See what kind of "wheels" you can create. Change the size. Design a new shape. Try a seatboard where you sit right on the blank. Use runners instead of wheels, and you'll have something for the snow. Try wheels on skis and you'll be able to cross-country ski on the sidewalk. Try a larger board for cement surfing.

There is no limit to what you can create, but remember, your invention may not work the first time. It may not work at all, but you'll learn something and have fun every time you try.

Glossary

Axles—the bar or rod on which a wheel turns.

Baseplate—a piece of metal on the truck that can be attached to the board.

Bit—the end of a drill that can be removed and replaced to make different-sized holes.

Bolt—a large screw that takes a nut.

Cross-country—hiking on skis.

Custom—made to order or specially made for someone.

Grain of wood—the direction in which the fibers of wood go.

Lacquer—a clear protective coating that can be dissolved with acetone or nail polish remover.

Motocross—riding an obstacle course on a bicycle.

Nut—a piece of metal with a threaded hole that screws onto the end of a screw or bolt.

Parallel—two things facing in the same direction the same distance apart at all points.

Pedestrian—someone walking rather than riding.

Penny—a term used with a number (threepenny), once used to tell the cost of nails and now used to indicate their size.

70

Perpendicular—two things at right angles to each other.

Sawhorse—a frame to hold wood while it is being sawed.

Sealer—a light protective coating usually made of oil or wax.

Truck—a unit that holds the wheels and can be attached to the board.

Urethane—a plastic material.

Varnish—a protective coating, thicker than lacquer, that can be dissolved with paint thinner.

Vehicle—something that moves on wheels.

Vise—something used to hold an object firmly.

Washer—a flat ring used between a screw and a nut to make a tighter fit.

Wedeling—small weaving turns usually done on skis.

Weighting and unweighting—lifting and lowering one's weight.

Metric Equivalents Chart

There are two main systems of measurement, the metric system and the imperial system. Most of the world uses the metric system (millimeters, centimeters, and meters). The United States has always used the imperial system (inches, feet, and yards). Soon the United States will use the metric system too. Some tools, screws, and bolts are already being measured in millimeters and centimeters. The charts below will help you to make the changeover.

1 inch	=	2.54 centimeters or 25.4 millimeters
1 foot	=	0.3048 meter
1 yard	=	0.9144 meter
1 millimeter	=	0.0394 inch
1 centimeter	=	0.3937 inch
1 meter	=	39.37 inches, 3.3 feet, or 1.1 yards

To use the metric system for the projects in this book, round off the numbers as shown in the following chart.

¼ inch	=	6 mm	4 inches	=	10 cm
⅜ inch	=	9 mm	10 inches	=	25 cm
½ inch	=	1 cm	1 foot	=	30 cm
1 inch	=	2.5 cm	2 feet	=	60 cm
2 inches	=	5 cm	1 yard	=	90 cm
3 inches	=	7.5 cm			

Index

About the Authors

Marilyn Gould is a teacher and occupational therapist who has taught at many levels, including early childhood education, elementary school, and high school. Her interests in young people, writing, and sports are all reflected in this book. Mrs. Gould lives with her husband in Newport Beach, California. The Goulds have two daughters and one son.

George Gould, who is Marilyn Gould's son, is a student at the University of California at San Diego. He is a skateboard enthusiast whose interests also include sports of all kinds and building things with his hands. He is especially well-qualified to write about building skateboards because, he says, "as a child I made skateboards and scooters for myself and every kid in the neighborhood."

This is the first book written by this mother-and-son team.